PROGRAMMING AMATEUR RADIOS WITH CHIRP

Ham Radio Setups Made Easy

BRIAN SCHELL

Programming Amateur Radios with CHIRP
Ham Radio Setups Made Easy

Written and designed by: Brian Schell
brian@brianschell.com

Version Date: July 18, 2018.

ISBN-10: 1720767262
ISBN-13: 978-1720767268

Printed in the USA of America

CONTENTS

INTRODUCTION

"Back in my day, radios came with a CW key and a frequency knob, and that was all we needed." —*Every Old Ham Curmudgeon*

OK, that's probably a *slight* exaggeration, but radios were a lot simpler in the "distant" past, when everything was station-to-station, and most communications were some flavor of HF. Nowadays, we have HF, VHF, UHF, digital modes, CW/SSB/AM/FM/MW, repeaters with offsets, repeaters with tones, User IDs, reflectors, talk groups, and who-knows-what coming down the road for next year. There are a lot of modes, tones, offsets, and little nit-picky details that have to go into our radios before we can even make a call.

On the other hand, radios, especially handheld models, continue to get smaller, lighter, and generally have fewer physical buttons and controls. This is good in that it keeps costs down, adds to durability, and allows for waterproofing, but it doesn't make programming them any easier. Fortunately, we have computers to handle all the programming stuff for us. Just hook the radio up to the computer, enter in all your information, transfer the frequency and channel data into the radio, and you're good to go. Sounds simple, doesn't it?

The problem is that radio manufacturers aren't necessarily good software designers. God forbid you try to program a radio on an Apple or Linux computer using manufacturer's software. None of the companies that make radios include software for these two very common operating systems. None. Even Windows users aren't in the clear. What happens when you try to use the software that came with your radio, when the radio was sold in 2007 and the software was made for Windows XP? Or maybe your radio came with a state-of-the-art serial port connector. That was great... back when all computers *had* serial ports; now, those are hard-to-find options or require an adaptor. And even though many ham operators enjoy tinkering with computers, there are many who still haven't embraced computers.

These things just aren't that simple.

Fortunately, there is CHIRP, a free, multi-platform software that works with a large number of common amateur radios.

If you're already comfortable with using programming software from RT Systems, or have no trouble working with whatever software that came with your radio, than maybe this book isn't for you. If you aren't good with computers, or you're having trouble with the basic process, don't know what all those columns mean, or are otherwise pulling your hair out trying to get your radio programmed, then this book *is* for you! I'll walk you through getting the software installed and set up, connecting your radio with an appropriate cable and communications port, reading template data from the radio, editing that data, and writing that data back out to the radio. CHIRP even has database tools for setting up local repeaters, National Calling frequencies, MURS, FRS, GMRS, and

Marine frequencies— you may not even need to look anything up!

What is CHIRP?

CHIRP is software used for programming frequencies, channels, tones, and other connection information that works with a large number of radios, on all major computer operating systems. It covers most features of popular radios, and it is completely open-source and free.

Why Use CHIRP?

One major company, RT Systems, has created really good radio programming software for years. RT Systems does a great job, and their software and cables work reliably and consistently. They are excellent. On the other hand, they're also quite expensive, ranging from $30 to $60 per radio for a cable and a CD. If you have five or six handheld radios (and they do seem to pile up quickly), this can start counting up fast. Some cheap radios, Baofeng for example, may even cost less than the programming cable and software.

Many modern radio manufacturers include software to program their radios. Some of these are top-notch tools, and some are really poorly-made software included almost as an afterthought. In the best of these cases, I wouldn't suggest that CHIRP is necessarily better than what the manufacturers make. Although in some cases, CHIRP beats them by a mile. In some other cases, if you lose the CD or buy a used radio, you may not get the software and need something to replace it.

One more reason is that you may want to use a single piece of software to program ALL your radios. Only CHIRP works on a large number of radios. If you need to "cut and paste" a range of settings from one radio to another, it's easy to do with a tool like CHIRP that supports all your radios.

ALTERNATIVES TO CHIRP

RT Systems

As mentioned in the previous section, these guys have almost become the default software supplier for programming handheld radios. Whatever radio you have, they have good, reliable software that supports your radio. Their cables are excellent, and they are available everywhere that sells radios, so they are easy to find. They are, however, more than a little expensive.

Manufacturer's Software

Often poorly-made, included as an afterthought, and rarely, if ever, updated, most manufacturers are not software developers. Even if they do create good software, once a radio is designed and shipped, they may never look at (or update) the software again. Computer operating systems and hardware change quickly, and the software quickly goes out of date. With no one to update the software, it often becomes useless with the very next OS upgrade.

Keypad Programming

This is feasible on some radios, especially if only a few frequencies or repeaters need to be entered, but it's never simple. If there are a large number of repeaters, or if you program multiple banks, or get involved with D-Star or other digital modes, this often become unrealistic. Some radios and situations are just too much to handle with a dozen little rubber keys. The computer makes it much easier.

HARDWARE REQUIREMENTS

Radio

Obviously, you'll need a radio. More specifically, you'll need a *supported* radio. There is a list included in the "Problems and Resources" section of this book that was current as of the publication date, but new radios are being added all the time. For the current list, take a look at `https://chirp.dan-planet.com/projects/chirp/wiki/Home`

Cable

If your radio came with a programming cable, then you are *probably* in good shape. Just use the one that came with your radio. Make sure that the included cable and software weren't made by RT Systems. Some manufacturers, rather than create their own software, will simply include the RT Software and cable with the radio. These kinds of cables *may* not work with CHIRP. Some will, some won't; it's hard to know without trying them, but there are reports of some RT cables that are not compatible with CHIRP.

Serial Port or USB Adaptors

This is easy to overlook if you've never programmed a radio before. Make sure your computer has the right kind of "plug" for the radio cable. Most newer computers have USB ports instead of the old-style serial/RS-232 ports that used to be ubiquitous, but many older radios only included the older-style cables. If this is the case, then you will need to go one of two routes:

1. If you are using a desktop computer that can accept plug-in cards (like PCI Cards), then you'll probably get the most benefit out of buying one of these and installing it inside your computer. If you do a lot of radio-to-computer interfacing stuff, sooner or later, you'll run into multiple uses for a "real" serial port on your computer. I've included a link for the **StarTech.com 2 Port PCI RS232 Serial Adapter Card with 16550 UART** in the Problems and Resources section, although there are many other brands and options. This is a very time-tested and stable technology; the only real trick is to make sure whatever you buy will actually fit in whatever card slot you have inside your machine.

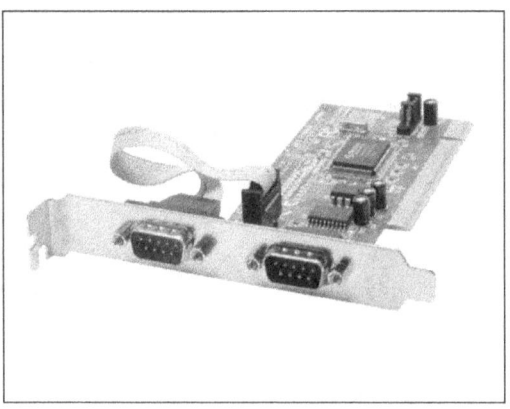

StarTech.com 2 Port PCI RS232 Serial Adapter Card with 16550 UART

2. If you are using a laptop or don't want to open up your computer, you can buy a USB-to-Serial adaptor at most computer shops or online. Just make sure that the adaptor is made with a genuine Prolific or FTDI chipset. There are numerous Chinese "clones" of this chip, and these often do

not work with radios. I have included a link to the **Tripp Lite Keyspan High-Speed USB to Serial Adapter, PC & Mac (USA-19HS)** in the Problems and Resources section.

Tripp Lite Keyspan High-Speed USB to Serial Adapter

INSTALLING CHIRP

CHIRP is easy to download and install, just go to `https://chirp.danplanet.-com/projects/chirp/wiki/Download` and click on the appropriate link for your computer:

CHIRP Download Page

Whether you choose Windows, Mac, or Linux, you'll wind up at a page that looks like this, with the version you need highlighted in green:

	Name	Last modified	Size	Description
	Parent Directory	-		
	Model_Support.html	19-May-2018 00:18	626K	
?	SHA1SUM	19-May-2018 00:18	411	
	Test_Report.html	19-May-2018 00:18	144K	
	chirp-daily-20180519-installer.exe	19-May-2018 00:18	11M	Windows Installer
	chirp-daily-20180519-win32.zip	19-May-2018 00:18	14M	
	chirp-daily-20180519.app.zip	19-May-2018 00:18	900K	MacOS Application
	chirp-daily-20180519.tar.gz	19-May-2018 00:18	737K	Source Tarball

Confused about what to download?

For more information, see the Download page! Quick summary:

- Windows users will want the **installer.exe** file
- MacOS users will want the **app.zip** file (and also need the runtime)
- Linux users may want the **tar.gz** file (but Ubuntu users should use the PPA)

Download the right version

Note that the filenames include the date of the most recent update. Always make sure you are downloading the most recent version of the software, as new features and radios are added all the time. The developers are very fast in squashing bugs and offering updates, so check back here for new versions regularly.

The installation instructions on the page are clear and *mostly* easy to follow, but here are a few clarifications if you need them:

Windows notes:

If you know specifically that you need a 32-bit version of the software, then download the version with "win32" in the filename; otherwise, download the "installer.exe" (64-bit) version.

Windows Installer

Mac notes:

For most Mac users, you will need to first download and install the "Python Runtime" file to install necessary tools that the CHIRP program will use. After that, download and install the MacOS installer.

If you are familiar with using the *Homebrew* package manager, then you could alternatively use that to install everything you need. Assuming you have *brew* (https://brew.sh) installed, you can simply type:

```
brew install tdsmith/ham/chirp
```

Linux

The web site lists the repository and apt-get commands to install the software. The ".gz" file in the download section is for source code, and not useful if you aren't going to develop or compile the code yourself.

One way or another, once you get the program installed, you should double-click on the icon as with any other app.

You might be surprised to see a mostly-blank screen when the
app starts, but this is normal:

Nothing to see here; an empty CHIRP screen.

Once you've got the software installed and running, it's
time to deal with the cable. Plug the cable into the radio,
leaving the radio turned off for now, and plug the other end
into your computer, or into your port adaptor if you need
one. With any luck, your system will "see" the cable and
everything will just work.

Let's see if you "won the lottery" or not:
1. Connect the cable and turn the radio on.
2. Pull down the "Radio" menu in CHIRP
3. Select "Download from Radio" from the menu.
4. A dialog box similar to the one below will appear.

Does the "Port" field contain a value? If it does, then your
system can see the radio; skip ahead to "Reading from the
Radio." If it's blank, pull down the "port" drop-down box and

see if *anything* is on that list. If there is nothing there, then your computer can't see the port, and we need to do a little more work. The following explains how to download a new serial port driver for Windows computers; Mac and Linux computers handle drivers differently, but probably won't have this problem.

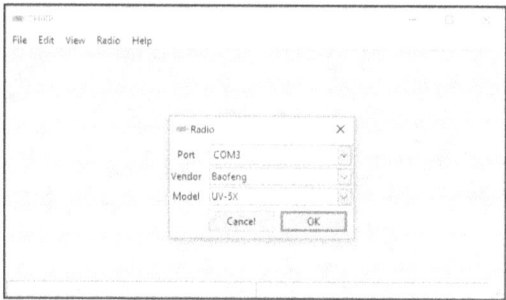

Choose COM port and radio manufacturer

5. In Windows, press Windows-X (The "Windows Key" on the keyboard along with the X) and choose "Device Manager" from the list of choices. The device manager looks like the window below. Make note of whatever your system says under "Ports (COM & LPT)." If it has some kind of error symbol next to it, then you are having a driver problem of some sort. This usually means that you are using a port adaptor that doesn't have a "genuine" Prolific chip in it. This is very common, and you'll need to download a special driver for it.

6. Find the driver at `http://www.miklor.-com/COM/UV_Drivers.php`. As of this writing, the driver is the one for **Prolific 3.2.0.0.** Follow the instruction on the website to install this driver. *Note that Windows 10 will overwrite this driver with newer versions of its own serial port driver with any major update—* the driver we are downloading and using is **old**, but it's the only one that works. You may need to

install it again after a major Windows update; just save the installer file and run it again if you have trouble in the future.

7. After the correct driver is installed, the "problem icon" should go away, showing something like the following. You MAY need to reboot your system for the changes to take effect. You need one of the lines under "Ports (COM & LPT)" to show a "COM" port. The one my system shows below is **COM3**.

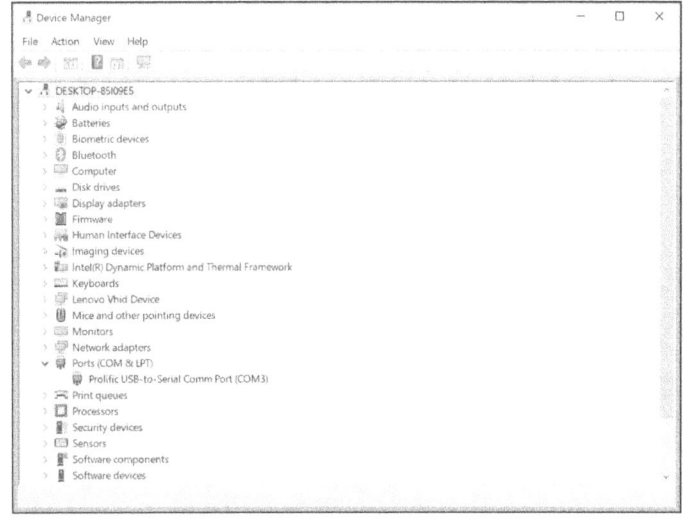

Windows Device Manager with COM port 3 showing

Again, Mac and Linux should not have the same problems with drivers that Windows has. If you *do* have a problem with those operating systems, look for an updated driver on the web and install it. Obviously, the process will be different than with Windows, but the concepts are the same.

THE MAIN PROCESS OVERVIEW

Generally, every action with CHIRP follows a three-step process:

1. Read data in from the radio (or load from disk)
2. Change things in the software
3. Write data out to the radio (or save to disk)

Each of these steps will be covered in detail in the next three chapters.

Optionally, you can also save data to the hard drive for editing and re-use at a later time. It's possible to have multiple data files and load them in as necessary. For example, I have a data file containing local stations here in Flint, where I live, but I also visit family in Dayton and Lansing and visit friends in Cincinnati. I have setups for all four cities, and when I know I'm traveling, I just load those cities into my radio.

RECOMMENDED FIRST-TIME PROCEDURE OVERVIEW

Again, we'll go over all this in detail in the next chapter, but here's the process above in a little more detail:

1. Use the "Download from Radio" function to create a template for your radio. The resulting file may or may not contain useful memory information at this time, but the file will be prepared and formatted for your specific radio.
2. This "radio dump" now becomes your working file. It's probably a good idea to save it now.
3. Next, you can open a stock configuration file containing the preset "channels" you want. Alternatively, you can use the "Import from Data Source" or "Query Data Source" to import repeaters and other standardized frequencies. Or if you know exactly what you want, you can enter repeaters and frequencies manually.
4. Edit, copy, paste, delete, and otherwise manipulate the memory channels as you see fit.
5. Save your completed file, perhaps with a new file name if you have changed things significantly.
6. "Upload to Radio" to write out your information to the radio.
7. Test the radio, and if necessary, go back to a prior step to fix things if it's not perfect.

Of course, after going through all this the first time, you don't necessarily have to download from the radio again if you prefer to load your configuration file from wherever you saved it.

READING FROM THE RADIO

Note: You MUST do the following steps, even if your radio is new and "empty." By reading from the radio, the fields that the radio supports are identified to the CHIRP software. Not every radio offers every feature, so this is where CHIRP "learns" what your radio can do.

OK, so your software is installed and can "see" a serial port. Pull down the "Radio" menu and choose "Download from Radio" to get the dialog box shown below:

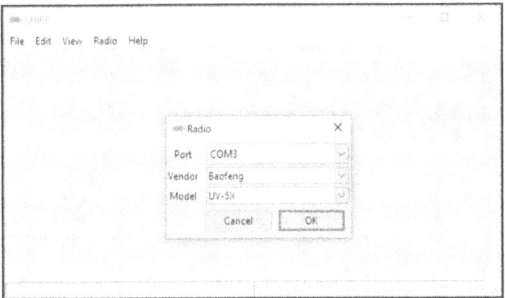

Choose COM port and radio manufacturer

As the dialog box asks, select whatever port your radio is

plugged into, then choose the manufacturer and model number of your radio. The model numbers may not match precisely; for example, for this chapter, I am using the Baofeng UV-5R radio, since it's extremely common and popular. CHIRP doesn't have a listing for that radio in the drop-down list, but it does show up in the list of supported radios. What's up? There is a listing for "UV-5X" radios. Baofeng has a number of radios that all start with "UV-5Something" and the "X" is simply a wildcard/placeholder. Your radio may use some similar kind of wildcard, so if it's not listed exactly, try something close and see what happens— some experimenting may be in order.

Anyway, once you have made the selections for your radio, click "OK." A scary-looking warning may pop up:

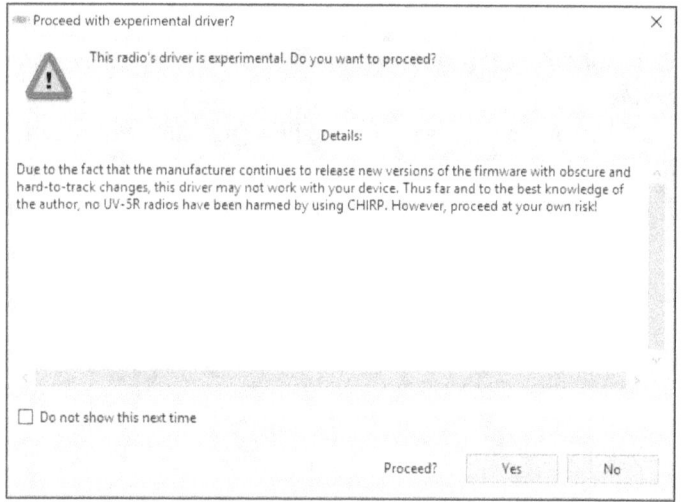

Experimental Driver

It says this for all the radios, so don't worry much about it. After you've read it once, it'll get old fast, so you may want to click "Do not show this next time." Then click "Proceed."

Next, we'll get a little dialog with more instructions. Mine looks like the image below, but they are slightly different depending on the specific radio. Do what it says; usually it tells you to turn the radio off and then on again, make sure the cable is connected, and perhaps something about setting the volume.

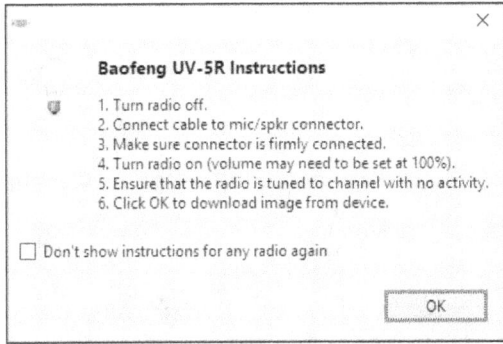

Baofeng UV-5R Instructions

1. Turn radio off.
2. Connect cable to mic/spkr connector.
3. Make sure connector is firmly connected.
4. Turn radio on (volume may need to be set at 100%).
5. Ensure that the radio is tuned to channel with no activity.
6. Click OK to download image from device.

☐ Don't show instructions for any radio again

OK

Last-minute Instructions

Click OK, and you'll get a little "cloning" dialog:

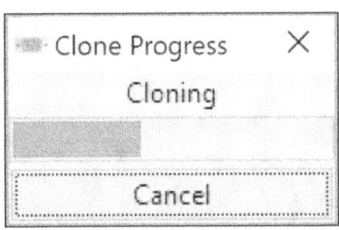

Clone Progress

Cloning

Cancel

Clone Progress Dialog

If you get an error at this point, there's probably still something wrong with your serial port or connection. Go through the steps again, and make sure to reboot after installing the serial port driver. If it still doesn't work, things get complicated. Check out the CHIRP support site for more personalized help, but especially take a look at your cables and drivers for problems.

If it does work, you will see a grid with various radio-related things in it, something like this:

Baofeng UV-5R Radio Programming Screen— Success!

Note that not all radios look exactly like this. Different radios have different features, and CHIRP only shows the features that you have access to. The above radio is a "cheap" Baofeng UV-5R. A more feature-packed radio, like the Icom ID-51A, shown below, has more tabs on the left and different columns across the top:

Icom ID-51A Radio Programming Screen

You first step before you change anything should probably
be to save your data. Pull down the file menu, choose "Save"
or "Save As..." as needed, and name the file something memorable. I generally use the radio model and my location—
"Baofeng UV5R Flint" or "ID51A Dayton" or something like
that. If your radio is brand-new, and there's either nothing in
the data file or generic default stuff, then just call it "Default
Settings" or something like that. No matter what you plan to
do with the radio, saving the default settings first is a good
way to make sure you can't screw things up too badly— you
can always reload what you started with. Once you've backed
up what *was* on the radio, then you can start changing things.

PROGRAMMING THE RADIO

You should always read in data from the radio before trying to program it for the first time. There are a couple of reasons for this:

1. It verifies that your COM port and connections are working and set up correctly.

2. It identifies what brand and model radio you have to CHIRP, so it can understand the capabilities of the radio. For example, if your radio doesn't do D-Star, you won't see the programming stuff for D-Star. If your radio doesn't have "banks" for memories, you won't see options for that. CHIRP only shows you the things you can use, not every possibility in existence.

3. In many cases, even a "factory fresh" radio will have something useful in those memories, and it doesn't hurt to keep a backup of those files before you go in and start modifying things.

MENUS

The menubar at the top of the CHIRP screen is where you perform various actions on the data files.

FILE MENU

This is a fairly standard menu with the usual **Save** and **Save As...** functions as well as the following more interesting items:

"New" Normally, you'd expect to start here, but this is **NOT** usually the way to begin. "New," in this case, means it will create a new blank GENERIC CSV file, which is something you may only use in certain specific cases. What you probably want to do is to begin with the "Download from Radio" function located under the "Radio" menu first. This will set up a template for your specific radio.

"**Import**" and "**Export**" CHIRP has its own internal file format, but if you desire, you can also save and load from a CSV (Comma-separated Values) file. These allow you to import information on specific memory locations from other files, which can be useful if you are copying similar information from one radio to another.

"**Open Stock Config**" This sets up a new GENERIC CSV file (not one that can be uploaded directly to your radio) in a new tab, containing all the frequencies for the chosen configuration. For example, "US Calling Frequencies" is one I usually import into all my radios, as this never really changes. It sets up a file with memory locations pre-configured for 6m (53.525 mHz), 2m (146.52 mHz), 220cm (223.5 mHz), and 70cm (446.0 mHz). These are common, easy to remember frequencies, but it's nice to be able to have them just "be there" automatically. Keep in mind that most handheld radios can't hand all four of those frequencies, so you may need to delete one or more of them before copy and pasting them into your radio file. It's also nice to have the whole list of Marine frequencies, railroads, MURS, and NOAA channels available just by loading a stock config. Play around with these, and see what all you can find that interests you.

Import from Stock Configurations

EDIT MENU

This menu contains the usual features like cut and paste, but the interesting ones here are:

"**Move Up**" and "**Move Down**." These two commands will re-order lines in your file as desired. If there's a station halfway down the list that you want to have at the top, just "Move Up" until that's where it lands. You can also swap two lines by selecting them both and choosing the "**Exchange**" menu choice.

"**Properties**" brings up a dialog box that lets you enter or change all the information for a selected line/memory location. You can also change all this data right in the main list by clicking on it and editing the data in column form. Either way works, it just depends on whether you want to do it in-line or in the dialog window.

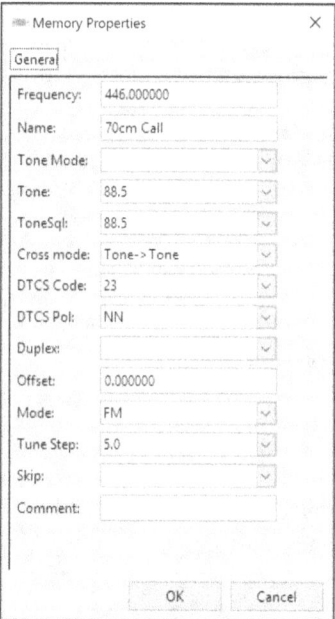

Properties Dialog for Memory Channel

VIEW MENU

"**Columns**" allows you to show or hide specific columns. You can see in some of my previous listings for the channels on the Baofeng and Icom radios that there are several "empty" columns in the center. If I wanted, I could simplify things by hiding those useless columns.

"**Language**" allows you to choose the language of all the menus and dialogs within CHIRP. Since you're reading this right now, it's a safe bet that you want to leave this set to "English" or "Auto." If there's a need to change languages, the option is there.

HELP MENU

"**Get Help Online**" simply brings up the CHIRP homepage and Wiki. There's a lot of good information here about what to do when things go wrong, so don't forget about this if you run into trouble at some point.

"**Report Statistics**" send information about your usage to the developer. If you want to help the developer create better software, leave this on.

"**Show Instructions**" allows you to hide some of the dialogs used when uploading and downloading to the radio. They don't hurt anything, and make it clear what you should be doing, so I always leave them on. This is optional, and completely up to you.

.   .   .

"**About**" shows the version date of CHIRP and the included libraries. This is useful if reporting problems to the developer or if you're unsure about how old the version of the software is that you are using.

RADIO MENU

This is the important one. This is where the real work gets done in CHIRP.

"**Download from Radio" and "Upload from Radio**" These will be thoroughly covered elsewhere, but are crucial to programming your radio.

"**Import from Data Source**" and "**Query from Data Source**" These are my favorite functions in CHIRP. I always used to have to scrounge the web for local repeaters or look in some kind of outdated book for local frequencies. Using the sources listed here, I can pull up repeater information about a location I am going to and be reasonably confident that at least some of them will work. Notice how I didn't claim they were foolproof? Repeaters come and go, often with no one updating the reference sites; that's one drawback of a hobby run by, well, *hobbyists*.

"Import from Data Source" allows you to check off frequencies and automatically merge or overwrite memory locations with the chosen channels. "Query Data Source" brings in the same information in a new tab where you can cut and paste or otherwise view and manipulate the information without affecting your existing files.

Either way, choose a source that works for you; some sites

require user names and passwords on their respective sites. RadioReference.com requires a paid membership (although it is a very good system). My personal preference is Reapeater-Book.com - They're free, and have very flexible search options.

The other options is, of course, to look up and enter repeater information manually, which we'll discuss in a later section. Importing them from an online system is the easiest, although perhaps not the most complete or accurate, method of doing this.

"**Channel Defaults**" allows you to choose which Band Plan you wish to follow on your radio. "North American Band Plan" is the default for me, but the IARU Regions and Australia are other allowed choices.

Choosing a Band Plan

EXAMPLE RADIO SETUP

OK, so now I'm going to show you how I use some of these tools. Let's walk through programming my Baofeng UV-5R radio, a particularly inexpensive and common radio that is more-or-less basic with functions that are common to most handheld radios. Assuming you've installed CHIRP, hooked up a cable, and downloaded a template file from your radio, you'll now have a blank radio file to work with. If you already have some channels set up that you want to keep, that's fine; you can just append other things at the end. I'm going to start by Downloading From Radio, then deleting all the memories that are already programmed. This way I can "Start fresh."

If you want to clear out the memories on your radio, all you have to do is click on any one of the memories on the list, then pull down the EDIT menu, choose SELECT ALL, and then pull down the EDIT menu again and choose DELETE. This should clear out whatever is already on the list.

Here's my empty radio:

Empty Radio Template

My goal with this project is to program the radio with my local 2m and 70cm repeaters as well as the two national calling frequencies for those bands.

First, I want to put in my local repeaters. My radio can handle the 2 meter and 70 cm bands, so I will want those repeaters. I pull down the Radio menu and choose "Import from Data Source" and then choose "RepeaterBook," then "RepeaterBook Proximity Query."

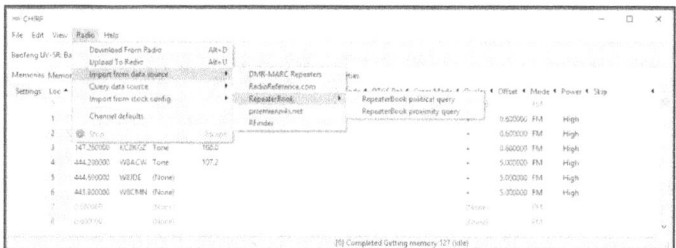

Digging into the Data Import functions

This will then pop up a dialog asking for Location, Distance, and Band. If I had instead chosen "RepeaterBook Political Query," I would be able to choose the State, County, and Band.

I put in my ZIP code, as well as the distance around that ZIP code (I want a 5-mile radius), and I'll start with the 2m band:

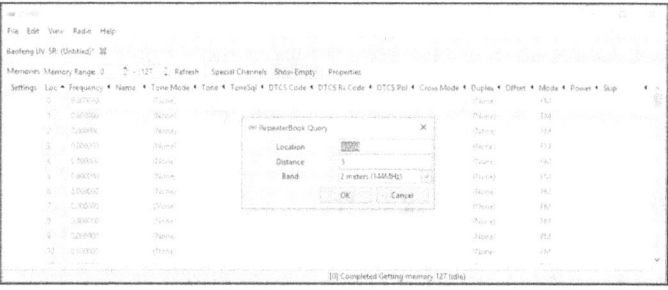

Repeaterbook Proximity Query

This brings up the three repeaters that are within 5 miles of ZIP code 48504:

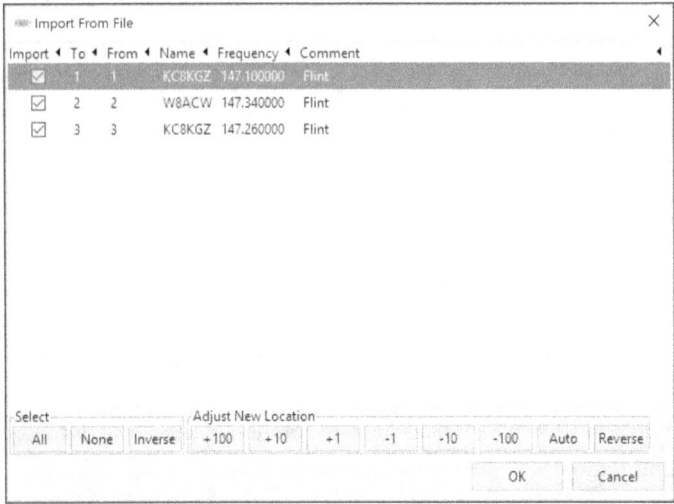

2m Band Repeaters near Flint, MI

Flint doesn't have much selection of repeaters, unfortunately. This screen has a few things to notice. The "Import" check box is either checked or not depending on whether we want to import it. I want all three of these, so I won't change anything. Next is "To" and "From." Now in this case, my radio working file is empty, so I want to start at memory location 1, so I don't need to change anything here either. If, however, I wanted to import these at some position OTHER than 1, I could use the buttons at the bottom of the window to add, subtract, or otherwise manipulate the location where these would be imported (more later). Then we see the name of the repeater, the frequency, and the city (as a comment). I want all these, so I'll just hit "OK."

Next, I want to import the repeaters for the 70cm band. I'll go back and do the "RepeaterBook Proximity Query" option again, this time changing only the band information. I see the following:

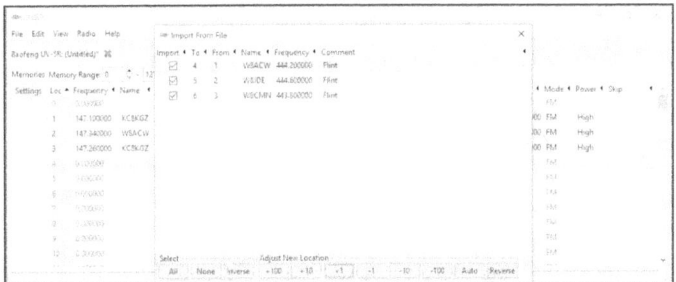

70cm Band Repeaters neat Flint, MI

Again, I want all three of these. This time, however, I already have stations programmed in at memory locations 1, 2 and 3 (you can see them behind the top window). I want these new memories to begin at location 4, so click the "+1" button 3 times to get what you see above in the "To" field. When that's good, I'll hit "OK" and they will import at that location.

As an aside, instead of importing the 2m and 70cm bands separately, I could have imported "All" bands and then deleted the ones that aren't compatible with my radio. Whichever way you think is easier for your location works fine.

For my third import, I now want the national calling frequencies for the USA. These don't come from Repeater-Book, but instead are in one of CHIRP's "Stock Config" files. Pull down the "Radio" menu, choose "Import from Stock Config," and then choose "US Calling Frequencies." Four different frequencies appear, but my little handheld radio only works on 2m and 70cm, so I don't need the other two. I will now uncheck the two I can't use. When I hit the "+1" button like before, I get a gap between numbers— it wants to import these as memory locations 8 and 10 (see below). What I can do to fix this is click on "auto." This will ignore the unchecked lines and renumber the two lines with checkboxes as "o" and "1." That's still wrong, but now I can hit "+1" until it imports them into location 7 and 8. It's a little fiddly, but we can get what we want eventually.

Importing Calling Frequencies

And then it's just a matter of writing the file back out to the radio. This is the "easy" way to import repeaters and frequencies and set up your radio. Of course, you can enter all these things manually as well. That would be done in the Memory Editor window, which we'll cover next.

THE MEMORY EDITOR SCREEN

The Memory Editor Window is the main window of CHIRP. Each line shows a memory location (or channel), and working across the columns, you can set all the options for each memory.

Memory Editor Screen

CHIRP MENU

At the top of the window is the usual menu bar with File, Edit, View, Radio, and Help. This was discussed in the section on Menus.

FILE TABS

Directly beneath that are one or more tabs that allow you to have more than one file open at a time. In the screenshot above, I have an *Icom ID-51*, a *Baofeng UV-5R*, and a *generic file containing US Marine VHF channels*, with the Marine frequencies being the currently selected tab. This makes copying and pasting between radio setups easy. You have the option of clicking on one tab, then right-clicking and COPYing the channel onto your clipboard. Then you can click on a new tab or empty memory channel and PASTE it in. If one radio allows some settings that the other doesn't, you may need to do some adjustments, but overall, it works very well.

For the three tabs open in the screenshot above, I can copy memories from the Marine band to either the Icom or the Baofeng tabs. I can copy the local repeaters from the Baofeng to the Icom just fine as well. On the other hand, you have to pay attention to what the radios are capably of doing: It will also allow me to copy the D-Star memories from the

Icom to the Baofeng radio, but those aren't going to work properly since the Baofeng doesn't support D-Star.

MEMORY EDITOR OPTIONS

The line beneath the file tabs contains several interesting items.

First, If you have a very long list of frequencies, you also limit the number of channels displayed by entering the memory range in the line below the tabs. You can use this to narrow down your editing to a few channels or to jump to a specific place in the list.

Next are several "buttons": Refresh, Special Channels, Show Empty, and Properties.

Refresh: Redraws the editor screen if necessary.

Special Channels: On most radios, this does nothing. Depending on the specific radio, these may show locked-down channels, such as in the 60-meter band.

Show Empty: Hides all the empty channels. If you have, for example, HAM stations in memory locations 1-9 and NOAA channels starting at memory location 100, then you probably don't want to see all the empty lines in the middle.

Properties: If you select a channel and then click on Properties, a dialog opens, allowing easy changes to the channel's settings. You can also just click on the value in the Memory Editor Window and edit settings that way; it's your choice.

MEMORIES TAB

Loc

The location in the radio's memory, also sometimes called a memory channel. How many of these you can use depends on your radio.

. . .

Frequency

This is the frequency associated with the memory channel. For a simplex channel, this is both the receive and transmit frequency, whereas a duplex channel (for example, a repeater) will often have an offset that makes the receive and transmit frequencies different.

Name

This is the human-readable name for the memory channel. This can be a callsign, a physical location, a club name, or whatever you want to put in to describe the frequency.

Tone Mode

Does the repeater require a special tone to access it, and if so, what kind of tone? Options are **None**, **TSQL**, **DTCS**, and **Cross**.

Tone

If a "Plain" sub-audible tone is required to activate the repeater, what is the frequency of the tone? This can be set (from a pull-down menu) to **NONE** or may range from **67.0 to 254.1**.

ToneSQL

If a "Tone Squelch" tone is required, what is the frequency of the TSQL? These range (from a pull-down menu) from 67.0 to 254.1

. . .

DTCS Code

If a "DCTS" tone is required to transmit, what is the DCS code? These range (from a pull-down menu) from 23 to 754.

DTCS Rx Code

If a "DCTS" tone is required to receive, what is the DCS code? These range (from a pull-down menu) from 23 to 754.

DTCS Pol

This is the Polarity of the DCS code. These can be one of (from a pull-down menu) the following: **NN**, **RN**, **NR**, or **RR**. The first character in these pairs represents Transmit Polarity, and the second is the Receive Polarity; the letters stand for either **N**ormal or **R**everse polarity.

Cross Mode

This means that the radio uses some more complex combination of Tones, DCS codes, and squelch tones. These can be one of (from a pull-down menu) the following: **Tone->Tone**, **Tone->DTCS**, **DTCS->Tone**, **->Tone**, **->DTCS**, **DTCS->**, or **DTCS->DTCS**.

Duplex

Tells the radio which direction the transmit frequency is offset from the receive frequency. This can be +, -, **split**, **Off**, or **None**.

Offset

For use with a duplex repeater, this is the value to add or subtract from the receive frequency to calculate the transmit frequency. For example, if the value in the frequency column is 145.11, the duplex is set to "minus," and the offset is set to 0.60000, then the radio will transmit at (145.11-.6) = 144.61 MHz.

Mode

Depending on what the radio supports, the drop-down menu lets you choose from **FM**, **WFM** (Wide FM), **NFM** (Narrow FM), **AM**, and **DV** (Digital Voice, a.k.a. D-Star).

Power

Sets the transmit power, depending on the radio. Typical options are **High** or **Low**.

Tune Step

This determines the increase of frequency that occurs when the knob is turned.

Skip

Choices: **None**, **S**, or **P**. When using the radio's Scan feature, this can make the radio **S**kip the channel or to make it a **P**riority.

Comment

A place for a short comment. How much space is available depends on the radio.

· · ·

URCALL

This is a field used only with D-Star compatible radios. URCALL is usually a "command" such as CQCQCQ, L (link), U (unlink), E (echo), I (Information), REF001AL (Link To Reflector 1A), or something along those lines. These commands are programmed in the "Your Callsign" listing in the D-STAR Tab on the left sidebar of the CHIRP Memory Editor Window.

RPT1CALL

This is another field used only with D-Star compatible radios. RPT1CALL is usually a repeater gateway such as W8HEQ or something along those lines. These repeaters are programmed in the "Repeater Callsign" listing in the D-STAR Tab on the left sidebar of the CHIRP Memory Editor Window.

RPT2CALL

This is another field used only with D-Star compatible radios. Similar to RPT1CALL, this refers to one of the repeaters entered in the "Repeater Callsign" listing in the D-STAR Tab on the left sidebar of the CHIRP Memory Editor Window. This is most often used as the "destination" in a gateway command.

D-STAR TAB

This book isn't about the specifics of D-Star programming, as it's a fairly advanced procedure. I would recommend my other book "D-Star for Beginners," available wherever you picked up this book.

. . .

Your Callsign

Despite what it sound like, this is actually a list of commands that can be sent to a D-Star Repeater.

Repeater Callsign

This is a list of repeaters and gateways.

My Callsign

This is exactly what it sounds like: Your Callsign. This is so the D-Star system can identify you.

TABS

There are a number of tabs on the left side of the CHIRP screen. How many depends on what kind of radio you have.

Mostly, these tabs do not involve specific memory locations or channels, but instead focus on features of the radio itself that can be adjusted through software. Some radios are easily configurable this way, others are not.

On my Baofeng UV-5R, there are only two: "Memories" and "Settings", while my Icom ID-51A has those plus "Banks," "Bank Names," and "D-STAR." What you see depends on your radio's capabilities, so there is no way I can explain all options available here.

To show a sample of what you may see, I have included all the screens that download from my UV-5R. Again, other radios will *significantly* vary, but you can see the kind of settings that are available. I will not go through each line, but will point out the noteworthy items:

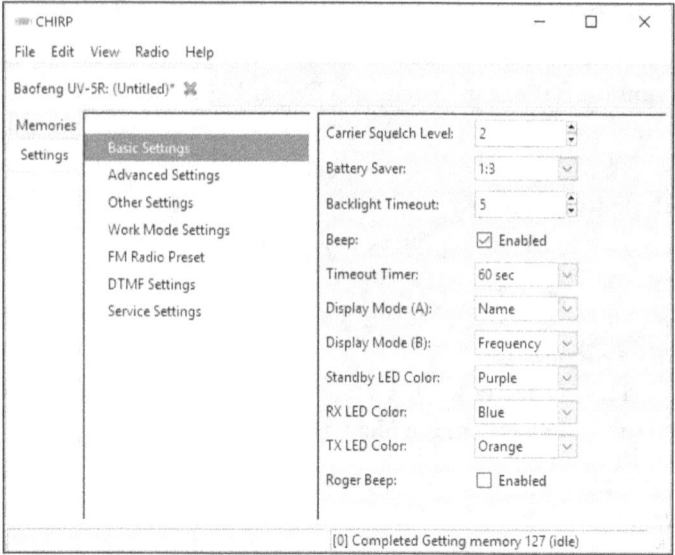

Basic Settings (Baofeng UV-5R)

On the "Basic Settings" screen, you can set the LCD colors, squelch, and the type of information that shows up on for display mode, both VFO A and B show up as channel names, and with display mode B, both VFOs show only the frequency number.

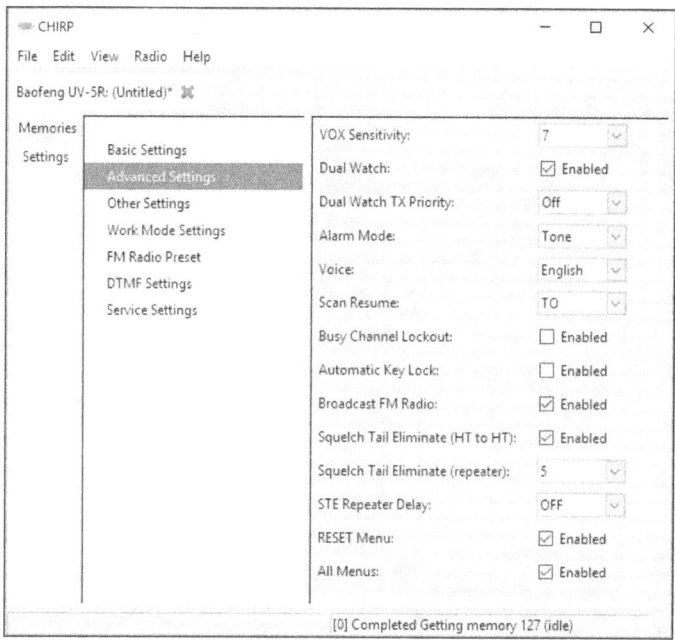

Advanced Settings (Baofeng UV-5R)

The "Advanced Settings" screen allows you to set the language, turn on single- or dual-watch features, and to turn on or off broadcast FM radio.

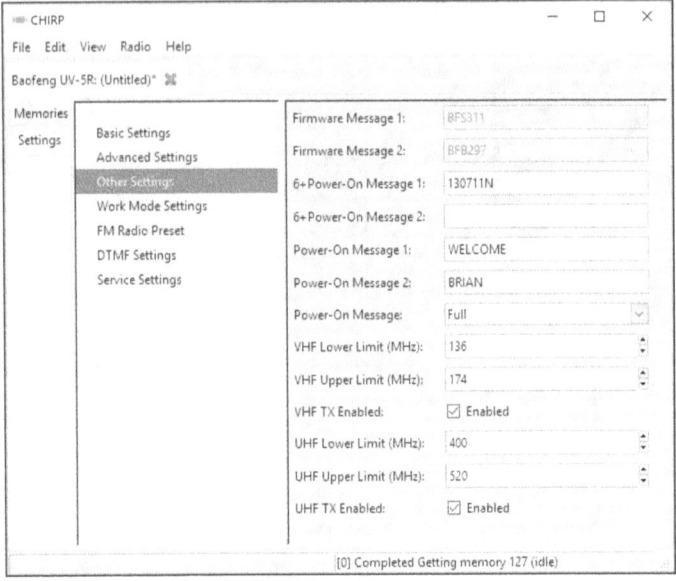

Other Settings (Baofeng UV-5R)

"Other Settings" include the text that displays when the radio is powered on, "Welcome Brian" and band limits.

The upper- and lower-limit settings lock out all frequencies that are not in the ranges shown. Be aware that the VHF and UHF upper- and lower-limits shown here are not legal to broadcast on. In the USA, transmitting on the 2m band is legal from 144.1 to 148.0 MHz, and the 70cm band is from 420 to 450 MHz. As you can see from the screenshot above, the radio is capable of a wider range of frequencies than your license will allow. In my case, I like to listen to off-band frequencies to see what I can find sometimes, and I only transmit on my pre-programmed repeaters or simplex channels, so I have chosen to leave these at the defaults. Depending on what you plan to do with the radio, it may be safer to enter in the actual lower and upper limits for your locale.

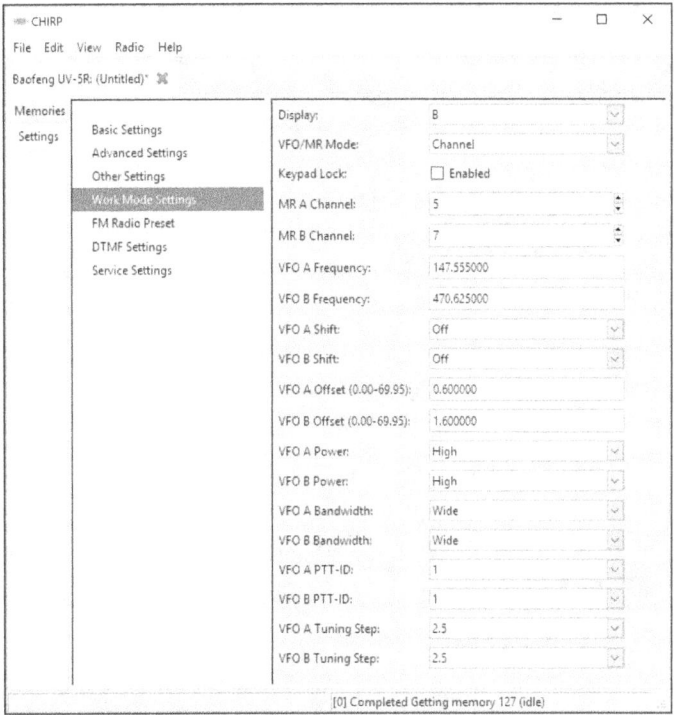

Work Mode Settings (Baofeng UV-5R)

The "Work Mode" settings allow for setting up default channels and/or frequencies, power levels, tuning steps, and other VFO options.

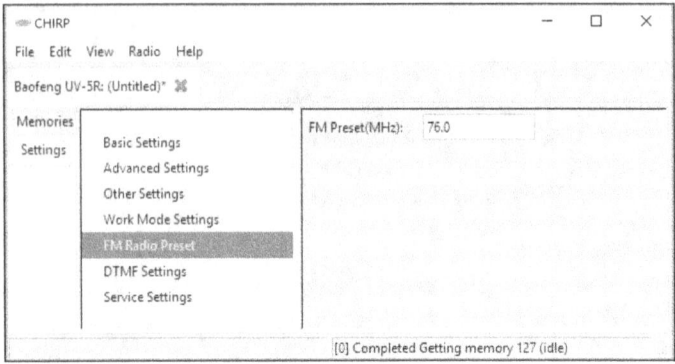

FM Radio Preset Settings (Baofeng UV-5R)

The "FM Radio Preset" screen only has one option. You can set the default frequency for turning on FM Radio mode. Put your favorite station in here, and your music is just one button-click away.

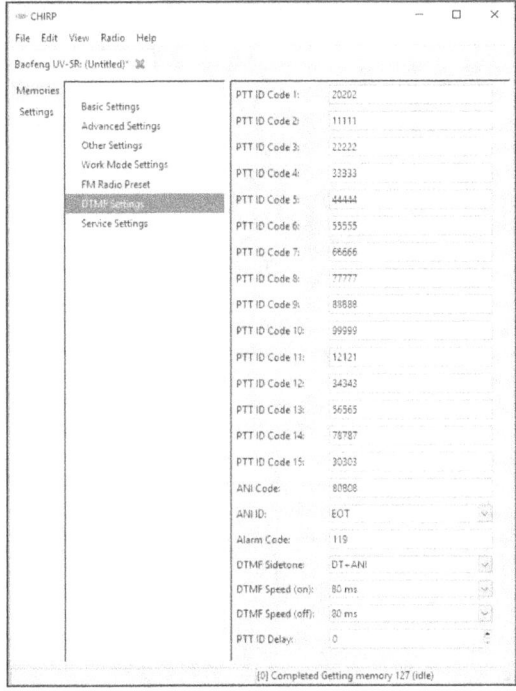

DTMF Settings (Baofeng UV-5R)

"DTMF Settings" is exactly what it sounds like. It gives you the option of programming/changing DTMF codes.

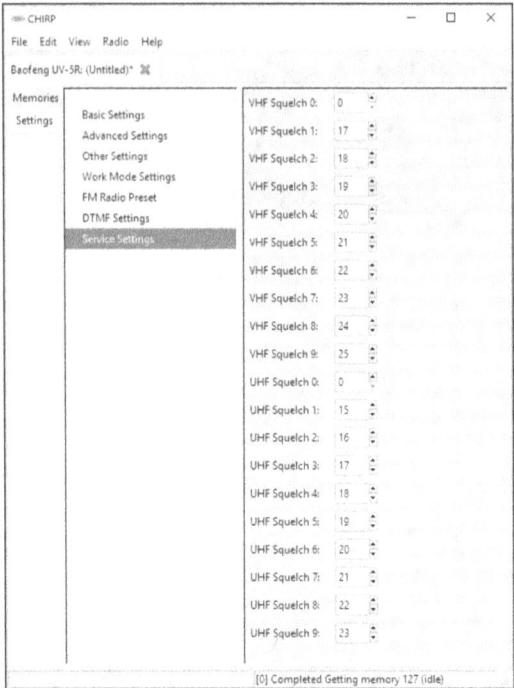

Service Settings (Baofeng UV-5R)

"Service Settings" allows you to customize and fine-tune squelch settings for both VHF and UHF squelch levels.

WRITING TO THE RADIO

OK, you've read in your data, you've added, deleted, or otherwise changed some data, now you need to get it back out of the computer and into the radio. If you were initially able to *read* from the radio, then writing back out to it shouldn't be any trouble.

Before you change anything on the radio, save your data. Pull down the file menu, choose "Save" or "Save As..." as needed, and name the file something memorable. I generally use the radio model and my location— "Baofeng UV5R Flint" or "ID51A Dayton" or something like that.

Writing to the radio is essentially done using the same steps as reading from the radio. Plug in your cable, turn the radio off then back on, pull down the "Radio" menu, and choose "Upload to radio."

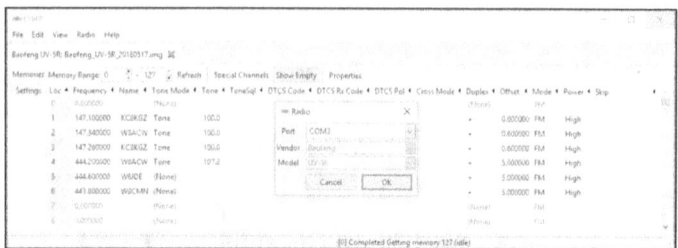

Upload to Radio Dialog

Just like when downloading from the radio, choose the port, vendor, and model of your radio, and then hit OK. As before, you should get some final instructions. Yours may vary depending on the type of radio you use:

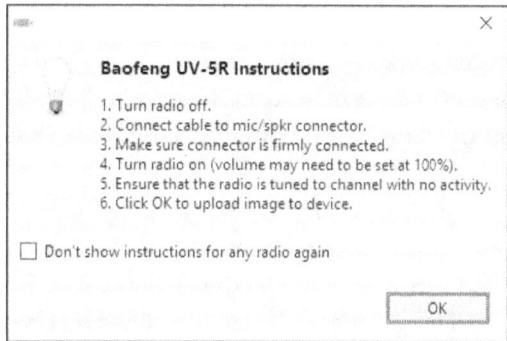

Baofeng UV-5R Instructions

1. Turn radio off.
2. Connect cable to mic/spkr connector.
3. Make sure connector is firmly connected.
4. Turn radio on (volume may need to be set at 100%).
5. Ensure that the radio is tuned to channel with no activity.
6. Click OK to upload image to device.

☐ Don't show instructions for any radio again

OK

Radio Uploading Instructions

When you are ready, click OK, and the computer will write data to the radio. Once it's finished, turn off the radio, unplug the cable, then turn the radio back on and test your changes. If something doesn't work the way you intended, change it in the software and write to the radio again, repeating as many times as necessary to get everything just the way you want it.

PROBLEMS AND LIMITATIONS

Solving problems

CHIRP is a work of love by its developers, offered for free to anyone who needs it. Compared to paying $40-$80 per radio for the next-best competitor, that's awesome! On the other hand, it's a mighty big project for just a handful of "hobbyists" to maintain. There are features that don't fully work yet, and occasionally something that used to work stops working. There are legal issues with reverse-engineering radios that bring up complications that go way beyond anything "computer guys" can deal with (this is one reason there is no DMR support in CHIRP).

CHIRP's official documentation is available at http://chirp.danplanet.com, but CHIRP doesn't have the best documentation (hence the reason I wrote this book).

Documentation:
https://chirp.danplanet.com/projects/chirp/

wiki/Documentation

FAQ:
https://chirp.danplanet.com/projects/chirp/
wiki/FAQ

Wiki:
https://chirp.danplanet.com/projects/chirp/
wiki/index

Bug Tracker:
https://chirp.danplanet.com/projects/chirp/
issues

Email List:
http://intrepid.danplanet.com/mail-
man/listinfo/chirp_users

Email List Archive:
http://intrepid.danplanet.com/pipermail/
chirp_users/

ADDITIONAL RESOURCES

StarTech.com 2 Port PCI RS232 Serial Adapter Card with 16550 UART
https://www.startech.com/Cards-Adapters/Se-
rial-Cards-Adapters/2-Port-16550-Serial-
PCI-Card~PCI2S550

Tripp Lite Keyspan High-Speed USB to Serial Adapter, PC & Mac (USA-19HS)

https://www.tripplite.com/keyspan-high-
speed-usb-to-serial-adapter~USA19HS

Using the Baofeng UV-5R as a Radio Scanner with CHIRP

https://oneguyoneblog.com/2018/02/25/scan-
ner-radio-pofung-baofeng-uv-5r-chirp/

Miklor CHIRP Page. Miklor specializes in Baofeng radios, but there are lots of good ideas here:
http://www.miklor.com/COM/UV_CHIRP.php

Miklor USB/Serial Drivers Page for Windows and Mac
http://www.miklor.com/COM/UV_Drivers.php

APPENDIX: SUPPORTED HARDWARE

The following list was copied from Chirp.Danplanet.com on May 27, 2018. Check the site for an updated listing of radios supported by CHIRP.

AnyTone
- AT-5888UV
- *Also includes the Intek HR-2040*
- *Also includes the Polmar DB-50M*
- *Also includes the Powerwerx DB-750*

Alinco
- DR-03T
- DR-06T
- DR135T
- DR235T
- DR435T
- DJ596T
- DJ175T

- DJ-G7EG

Arcshell

- AR-5
- AR-6

Baiston

- BST-2100 (use Baofeng BF-888)

Baofeng/Pofung

- F-11
- GT-3WP
- UV-3R
- UV-5R *and variants* (2 power levels)
- UV-6
- UV-6R
- UV-82/82L/82X
- *Also includes the GT-5*
- UV-82C
- UV-82HP/82DX/82HX (3 power levels)
- *Also includes the GT-5TP*
- UV-82WP
- *Also includes the UV-5RWP*
- UV-B5/B6
- *Also includes the BF-V85*
- BF-666S/777S/888S
- *Also includes the GT-1*
- BF-A58
- BF-F8HP (3 power levels)
- *Also includes the BF-F9V2+*
- *Also includes the GT-3TP*

- *Also includes the UV-5R PLUS*
- *Also includes the UV-5RHP*
- *Also includes the UV-5RTP*
- *Also includes the UV-5R7W*
- BF-T1 *(Same as Baofeng MINI or BF-9100 handhelds)*

Baojie

- BJ-218 *(Variant of Luiton LT-725uv)*
- BJ-UV55
- BJ-9900

BTECH

- GMRS-V1
- UV-2501
- UV-2501+220
- UV-25X2
- UV-25X4
- UV-5001
- UV-50X2
- UV-50X3
- UV-5X3

Feidaxin

- FD-150A
- FD-160A
- FD-268A
- FD-268B
- FD-288A
- FD-288B
- FD-450A
- FD-460A

. . .

Hesenate

- BJ-218 *(Variant of Luiton LT-725uv)*

HobbyPCB

- RS-UV3

Icom

- IC-80AD
- IC-2820H
- ID-800H
- ID-880H
- IC-208H
- IC-91/92AD
- IC-V/U82
- ID-RPx000V/RP2x
- IC-2100H
- IC-2200H
- IC-2300H
- IC-2720H
- IC-P7
- IC-T70
- IC-T7H
- IC-T8A
- IC-Q7A
- IC-W32A
- IC-746
- IC-7200
- IC-7100
- IC-7000
- ID-31A

• ID-51A

Intek
• KT-980HP

Jetstream
• JT220M
• JT270M
• JT2705M *(Variant of Waccom Mini 8900)*

Juentai
• JT-6188 Mini *(Variant of QYT KT8900)*
• JT-6188 Plus *(Variant of Waccom Mini 8900)*

Kenwood
• TH-D7A/G
• TH-D72
• TH-F6A
• TH-F7E
• TH-G71A
• TH-K2
• TK-260/270/272/278
• TK-260G/270G/272G/278G
• TK-360/370/372/378
• TK-360G/370G/372G/378G/388G
• TK-760/762/768
• TK-760G/762G/768G
• TK-860/862/868
• TK-860G/862G/868G
• TK-7102/8102/7108/8108

- TM-271A
- TM-281A
- TM-D700
- TM-D710
- TM-G707
- TM-V7A
- TM-V71A
- TS-2000

KYD

- NC-630A

Leixen

- VV-898
- VV-898S/898E

LUITON

- LT-316 *(Variant of Retevis RT22)*
- LT-580
- LT-588UV *(Variant of QYT KT8900)*
- LT-725UV
- LT-898UV *(Variant of Leixen VV-898)*

MTC

- UV-5R-3

Puxing

- PX-2R (UHF)
- PX-777

. . .

QYT

- KT-UV980 *(Variant of Waccom Mini 8900)*
- KT8900 *(same as KT-8900)*
- KT8900R
- KT7900D
- KT8900D

Radtel

- T18

Retevis

- H-777 (use Baofeng BF-888)
- RT-5R/5RV *(Variant of Baofeng UV-5R)*
- RT-B6 (use Baofeng UV-B5)
- RT1
- RT21
- RT22
- RT23
- RT26
- RT5 with 2 power levels *(Variant of Baofeng UV-5R)*
- RT5 with 3 power levels *(Variant of Baofeng BF-F8HP)*
- RT6

Rugged Radios

- RH5R *(Variant of Baofeng UV-5R)*
- RH5X *(Variant of Baofeng BF-A58)*

Sainsonic

- GT-890 *(Variant of QYT KT8900)*

Surecom

- KT8900D *(Variant of QYT KT7900D)*

TDXone

- TD-Q8A

TYT

- TH-UV3R
- TH-UVF1
- TH-7800
- TH-9000
- TH-9800

Yaesu

- FT-1D
- FT-50R
- FT-60R
- FT-70D
- FT-90R
- FT-817/ND
- FT-857/D
- FT-897
- FT-1802M
- FT-2800M
- FT-1900R/2900M
- FT-7800R/7900R
- FT-8800R
- FT-8900R

- FTM-350R
- VX-170
- VX-2R
- VX-3R
- VX-5R
- VX-6R
- VX-7R
- VX-8R

WACCOM
- MINI-8900

WLN
- KD-C1 *(Variant of Retevis RT22)*

Wouxun
- KG-UVD1P/UV2D/UV3D
- KG-UV6D/UV6X
- KG-UV8D
- KG-UV8D Plus
- KG-816/818

ZASTONE
- BJ-218 *(Variant of Luiton LT-725uv)*
- MP-300 *(Variant of QYT KT8900)*
- ZT-X6 *(Variant of Retevis RT22)*

CONCLUSION

And that's it. CHIRP in itself is not an especially-complicated program to use, but there are a large number of settings on each radio, and it needs to be able to manipulate all of them. Not to mention the ability to reliably connect to and communicate with all those different radio types. The truly outstanding part of this is that it's all done by volunteers and hobbyists and made available for free.

If, in the process of programming your radio, you run across a bug in CHIRP that you can't attribute to your own mistake (or mine), be sure to let the developers know about it — that's how they make the software better. While we're on that topic, if you've found something that *I* could have gone into with more detail or explained more clearly, let me know about it. Eventually, I'll put out a revised or updated edition, and your input can only help.

Good luck on the airwaves!

ABOUT THE AUTHOR

Brian Schell (KD8OTD) is a former College IT Instructor who has an extensive background in computers dating back to the 1980s. Currently, he writes on a wide array of topics from computers, to world religions, to ham radio, and even releases the occasional short horror tale.

He'd love to hear your stories of success and failure with programming your radio. If there's something you would like to see in a future edition of the book, or otherwise have suggestions, please drop him a note. Contact him at:

Web: http://BrianSchell.com
Email: brian@brianschell.com

twitter.com/BrianSchell

facebook.com/Brian.Schell

instagram.com/brian_schell

pinterest.com/brianschell

STAY UP TO DATE!

Join my email update list— There's NO weekly SPAM or filler material, only announcements of new books or major updates.

http://brianschell.com/list/

HELP ME!

CONTACT THE AUTHOR

If you have a suggestion or find a mistake, email me about it, and I'll get it into the next edition of the book. Got a gripe, complaint, question, or just adoring fan mail? Same thing!

LEAVE A REVIEW

If this book helped you, please leave a review where you purchased this book. Reviews are the best way to help out!

SHARE WITH YOUR FRIENDS

Did you enjoy this book? Please use the buttons below to spread the word to your friends and followers.

ALSO BY BRIAN SCHELL

Amateur Radio

• D-Star for Beginners

• Echolink for Beginners

• DMR for Beginners Using the Tytera MD-380

• SDR for Beginners with the SDRPlay

• OpenSPOT for Beginners

• Programming Amateur Radios with CHIRP

• FM Satellite Communications for Beginners

• Trunking Scanners for Beginners Using the Uniden TrunkTracker

Technology

• Going Chromebook: Living in the Cloud

• Going Text: Mastering the Power of the Command Line

• Going iPad: Ditching the Desktop

• DOS Today: Running Vintage MS-DOS Games and Apps on a Modern Computer

Old-Time Radio Listener's Guides

• OTR Listener's Guide to Dark Fantasy

The Five-Minute Buddhist Series

• The Five-Minute Buddhist

• The Five-Minute Buddhist Returns

• The Five-Minute Buddhist Meditates

• The Five-Minute Buddhist's Quick Start Guide to Buddhism

• Teaching and Learning in Japan: An English Teacher Abroad

Fiction with Kevin L. Knights:

• Tales to Make You Shiver

• Tales to Make You Shiver 2

• Random Acts of Cloning

• Jess and the Monsters